FARM B·A·B·I·E·S

A Grosset & Dunlap **ALL ABOARD BOOK™**

Library of Congress Cataloging-in-Publication Data

Rice, Ann.
Farm babies / by Ann Rice ; illustrated by Betina Ogden.
p. cm. — (A Grosset & Dunlap all aboard book)
1. Domestic animals—Infancy—Juvenile literature. [1. Domestic animals—Infancy.] I. Ogden, Betina, ill. II. Title. III. Series.
SF75.5.R53 1994
636'.07—dc20

93-26192
CIP
AC r93

ISBN 0-448-40212-2
A B C D E F G H I J

FARM B·A·B·I·E·S

By Ann Rice

Illustrated by Betina Ogden

Grosset & Dunlap, Publishers

This little kitten lives on a farm with her family.

She is just a baby. But she is not the only baby on the farm....

Baby rabbits are called kittens, too. They like to nibble on plants in the garden.

They didn't have any fur when they were born. But now they are as furry as their mother.

A baby donkey
is called a colt.
But this one's
long, pointy ears
look almost like
a bunny's. Hee-haw!

Over in the barn, a cow nuzzles her brand-new calf. She keeps her baby warm beside her in the hay.

Down in the pigpen, a mother pig lies down to feed her eight hungry piglets.

Each one has soft hair and a very curly tail.

The hen shows her chicks how to eat in the barnyard.

Peck, peck, peck. They pick up tiny seeds with their beaks, just like their mother.

A mother sheep watches her woolly lambs explore the meadow.

They like to smell the wildflowers. They also like to eat them!

A baby horse
is called a foal.
This one is racing
his mother across
the sunny pasture.
He learned to
stand on his long,
thin legs on the
day he was born.

Nearby in the pond, a mother duck is teaching her ducklings to swim.

When they are a little older, they will learn to fly.

Fuzzy little goslings will follow their mother wherever she goes.

But the goose doesn't mind. She likes to keep her eye on them.

Baby goats are called kids. They like to play by butting with their heads.

One day these brothers will have horns just like their father.

The mother dog keeps her five frisky puppies on the shady farmhouse porch. They will chew on anything. Even one another.

These puppies could play all day—and sometimes they do!

Inside the farmhouse, there is another baby.
Her big brother likes to rock her in her cradle.

Now where is the kitten going in such a hurry? Does she hear her mother calling her?

Yes! She knows it’s time for dinner—and a bath!

There are so many babies on the farm. And soon there will be more!